I0059506

Mobile Software Testing

Comprehensive Software Testing Guide
for Mobile Automation, Performance and
Security Testing

NARAYANAN PALANI

First published in 2014 by

BecomeShakespeare.com
Wordit Content Design & Editing Services Pvt Ltd
Newbridge Business Centre, C38/39,
Parinee Crescenzo Building, G Block,
Bandra Kurla Complex, Bandra East,
Mumbai 400 051, India
T: 91 22 33040620

Copyright © 2014 Narayanan Palani
All rights reserved. Any unauthorized reprint or use of this material is prohibited.
No part of this book may be reproduced or transmitted in any form or by any
means, electronic or mechanical, including photocopying, recording, or by any
information storage and retrieval system without express written permission from
the author/publisher.

Please do not participate in or encourage piracy of copyrighted materials in
violation of the author's rights. Purchase only authorized editions.

Copyright © 2014 Narayanan Palani

ISBN 978-93-83952-14-4

Photo Courtesy on Cover Page (from left to right):Satyadip Das, Santosh Muppidi
and Vimalaadhithan L

Preface

What this book is about?
This book describes how to test mobile applications and build automation, performance and security test knowledge implementation around test cycles. Practical knowledge on selective tools are included to benefit the automation testers to pick right tool for test automation. Collection of examples provided in certain chapters for better understanding over mobile testing implementation.

Who is this book for?
This book is intended to be of most help to software testing professionals who are working in mobile testing projects and interested to step into mobile testing assignments as beginners. The target audience for this book includes:
-potential and recent purchasers of test automation tools
-those who already have a test automation experience and interested to know mobile technology testing opportunities
-anyone who is building in house mobile test automation, security tools
-test tool vendors
-technical managers who want to insure that mobile testing projects provides benefits
-management consultants and domain consultants

Where this book can be suggested?
Computer Science students can be benefited from this book for the units related to software testing, automation testing, security testing and advanced mobile application testing.
Software Training Institutes can use this book as a basic guide for mobile devices testing and practice of automation tools for android applications.

Foreword

"Live as if you were to die tomorrow. Learn as if you were to live forever."

—Mahatma Gandhi

This is the Quote that suites Narayanan Palani, he is a kind of person who is hunger for knowledge. The success of his book "Advanced Test Strategy" greatly encourage him to write a second book.

Mobile application testing is one of the emerging domain for banking and financial services. I hope this book delivers the best automation and security testing concepts of mobile testing to software engineers and students in computer science.

I have known Narayanan right from his earlier days in banking assignments at Central London. He has great skills in testing technologies like orthogonal array, test matrix and automation script frameworks.

This book is in a simplified and easy to learn format for readers. It is a boon to all those who are into software testing and interested to take mobile testing opportunities. Considering Narayanan's caliber I expect him to come out with many more such useful books and publications. I wish him all the very best.

—Vimalaadhithan L
Test Manager
Polaris Software Lab, London, UK

Praise for the book- Mobile Software Testing:

"Mobile testing is increasingly complex on day by day due to the range of platforms,devices and innovations. Narayanan has articulated the complex mobile testing approach in simple terms with good references. I am sure, this book will enable QA community to pickup the latest developments in mobile testing arena and the tools available to deliver secured & quality product to the end users"

—Ponsailapathi Viswanathan
Vice President
Polaris Software Lab Limited UK
Northern Ireland Science Park
Belfast

Table of Contents

Table of Figures

Handwriting of Guru Raghavendra

Dedicated to
Sree Raghavendra Swamy
Sri Raghavendra Swamy Matha
Mantrayalam – 518 345
District: Kurnool
Andhra Pradesh, India

Introduction to Smart Phone Testing

An average consumer spends 127 minutes a day on the mobile phone and 49% of US customers are using smart phones[1]! Smart phones are used more than four times comparing to laptops and computers. So smart phones are the latest trend for the users to access emails, internet, store information, entertainment, install programs and games in one device which is used for mobile calls.

Smart phones provide new capabilities & applications at anywhere and anytime. So the need of having good quality applications with least possible error or issues is the advantage for any service provider! Android Testing needs the experienced test engineers in respective testing tools like Robotium which provides the challenge to Resource Allocation and Automation Framework Readiness. Mandatory trainings to test engineers needs to be concentrated mainly on the latest testing tools required for the upcoming mobile testing projects.

Mobile Software Testing

'An investigation conducted to provide stakeholders with information about the quality of the mobile products or service or the testable items as part of mobile testing'.

Testing smart phone applications and devices in various business conditions to prove that the quality standards are maintained and the application is reliable.

Emulators are the easy and cheap choice for any testing vendors. Example Brief on Test Closure Report: Testing is conducted on Retail online customer portals of ABC Stores Ltd and following are the number of defects reported during testing,

Major Defects: 3
Minor Defects: 35
Cosmetic Defects: 67
All the defects listed above are retested and closed at the end of test cycle.

Mobile Products

Variety of mobile devices are growing in the numbers across the countries and testing 'mobile applications' is the leading software testing trend in

IT Industry. Diversity of Mobile platforms like Android, IOS, BREW, Symbian, Windows phone and Blackberry are bringing up the real challenges to test engineers and knowing the scripting knowledge of particular mobile testing tool is become mandatory for any test experts. Over 400 mobile network operators are providing the services across the world and testing the customized services and applications provided by each vendor is a real challenge when there is no standard involved in testing.

Mobile Testing Industry

eMarketer expects 4.77 billion users across the varieties of smart phones in 2015 as per the latest survey[2]. 69.4% of population are expected to use phones by 2017. Minimum 5.13 billion customers are expected to use smart phone services in 2017. Hence mobile application testing is an emerging software testing practice across the IT companies as many of the upcoming projects are targeting mobile testing as a mandatory requirement for their products and services for smart phone services

Expected Growth of Testing Over Mobile

As an emerging sector of Information Technology, Software Testing started with Manual Testing Professionals and the target shifted to Automation Tests and Performance Tests later on. Security Testing, Localization Testing and Data Warehouse Testing are some of the niche testing practices in current technology region. Looking forward following are some of the expected niche testing sectors for testing experts:

1. Mobile User Acceptance Testing (Manual)
2. Mobile Automation Testing
3. Mobile Security Testing
4. Mobile Applications based Data Warehouse Testing

Defects and Failures

One of the expensive defect is requirement gaps i.e. unrecognized requirements which results in errors by program designer[3]

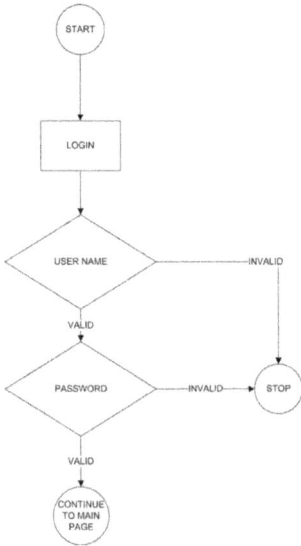

Figure 1 Input Flow

Example: Testable requirement says to test the Button 'Next', but when the button is clicked it is not going to next page. This information is not provided in the testable requirement. So this defect causes issues in the navigation of application.

Input Combinations

Minimum number of required tests to perform and complete during testing process needs set of input combinations.

Example:

Requirement: Login credentials needs to be tested

Number of possible input combinations:

User Name: Valid/Invalid
Password: Valid/Invalid

Total Combinations:
1. Valid user name with invalid password
2. Valid user name with valid password
3. Invalid User name with valid password
4. Invalid user name with invalid password

Roles

Software tester was a generic name used till 1980s and established to extensive roles and responsibilities over a period of time as follows:

1. Test Manager
2. Test Lead
3. Test Analyst/ Automation Test Analyst/Penetration Test Analyst/ Database Test Analyst
4. Test Designer
5. Automation Developer
6. Test Admin

Example: Rob Cooper is the Test Manager for Healthcare Domain Tests, handling three Test Leads for different streams. Each Test Lead are reporting to Rob and minimum 2 to 6 test analysts are reporting to Test Leads.

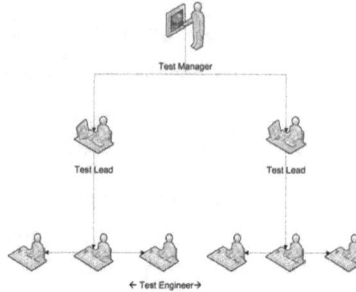

Figure 2 Software Testing Team Hierarchy

Sample Job Descriptions of Mobile Test Engineers in IT Industry

Role1 (Entry): Software Test Engineer-Mobile Applications

Responsibilities:

Experience in Mobile application testing. Product based application testing is an advantage. Good understanding of mobile architecture and mobile usage/internals are essential. Experience in iOS and Android application testing are preferred.

Expertise in writing test case, Preparing test plan and Identifying test scenarios according to the application. Good simulation skills, understanding user experience and out of box thinking.

Role2 (Mid Career): Senior Software Test Engineer-Mobile Applications

Responsibilities:

Primary Skills : Testing,IOS,Android,QA

Secondary Skills : Unix Shell Scripting,Oracle PL/SQL,

Personal Roles : Should have strong Knowledge in Device Anywhere and Perfecto Mobile. The test work primarily consists of testing mobile applications regarding application functionality, user interface, and testing from a backend perspective with regards to the data flowing between various IT backend servers systems.

Role3 (Management): Test Manager-Mobile Applications

Responsibilities:
4+ years of experience of leading a team with at least 1+ years of leading a mobile testing project Hands-on with testing mobile applications on mobility platforms such as Android, iOS, BlackBerry and Windows. Strong experience and hands-on with different phases of the STLC i.e. Test Planning, Design, Specification, Execution, Reporting and Defect Management Hands-on with mobility test infrastructure (physical devices / emulators / cloud based devices & connectivity). Exposure to PerfectoMobile or Device Anywhere is a plus. Hands-on with one or more Test Automation Tools for Mobility such as Robotium, JamoSolutions, Experitest SeeTest, PerfectoMobile or Device Anywhere. Strong exposure to Test Automation implementation with one of the test automation tools such as HP QTP, IBM RFT, MS VSTS or Selenium. Knowledge of performance testing will be a plus Strong exposure to Agile Methodology Hands-on with test effort estimations and planning Adaptability to delivery challenges Excellent communication skills Excellent people management and inter-personal skills.

Mobile Testing Process

Testing functionality, usability, reliability and it's consistency are the primary purpose to regularize the mobile testing process.

Unlike product testing, mobile testing needs special technical skills to understand the platform dependent user flow of functionalities. Also it is important to have scripting background and set of automation frameworks to regularize the testing process.

It is advised to follow STLC (Software Testing Life Cycle) approach with specific Mobile Test Activities included in each phase.

E.g.: Testing Android OS (Operating System) based mobile for email services need additional of Test Script/Tool Installation during Test Execution Phase.

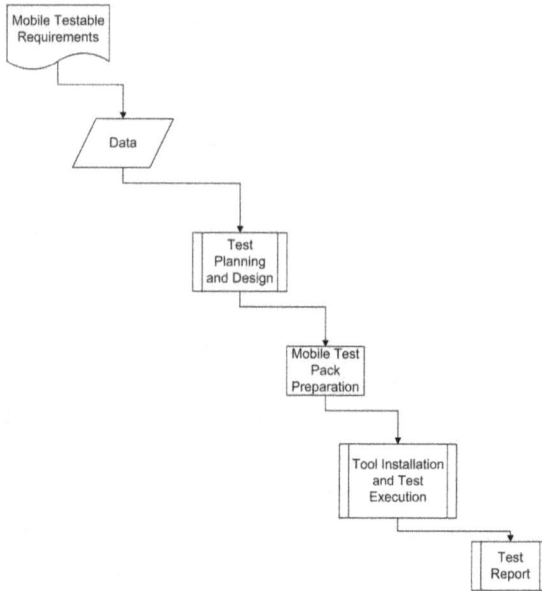

Figure 3 Mobile Software Testing Life Cycle

Mobile Testable Requirements:

Mobile usability and functionality based requirements are drafted in test requirements phase with the help of business and SMEs (Subject Matter Experts). Additional changes of requirements over the later stages are added as 'Change Requests' to the requirements.

Device specific guidelines, text entry, finger sized targets, work flow, consistency, call handling method, response times can be some of the important checkpoints while arriving at the testable requirements of usability testing for smart phones.

Network, Battery requirements, Volume, Rendering are some of the high priority checkpoints while arriving at the testable requirements of mobile performance testing.

Operating System specifications, Device and device driver requirements, Screen resolution and clarity requirements are some of the high priority checkpoints for deriving testable requirements of compatibility testing

Portability to different mobile devices for data exchange, Connectivity feasibility between computers and tablets are some of the key checkpoints for deriving testable requirements for synchronization testing.

E.g:

HellowMobileApp should respond the chat messages in 2 second time

Mobile applications are dynamic and competing with products available over the market is considerably important.

So defect free product is the need of customers for better response.

Good requirement of mobile applications should be clear about the functionality and user experience, understandable, complete in nature, reasonably detailed when listing the features and need, correct, consistent and unambiguous.

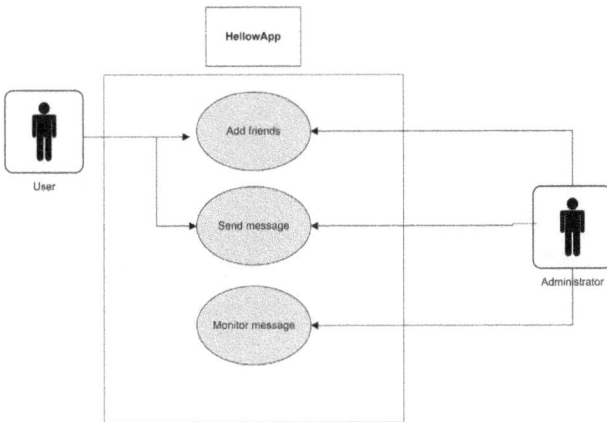

Figure 4 E.g Sample use case for a mobile application

Use Cases:

Detailed way of expressing requirements based on the perspective of users outside of the mobile devices.

They capture the uses of the mobile devices in terms of achieving goals or value for someone or something outside the devices and applications. That someone or something outside the mobile applications is called an actor.

Test Data Preparation

Complete test inputs and necessary test data for test execution needs to be collected as part of test data preparation. E.g. username and password required to login to the mobile application. Preparing different data for different test inputs and maintaining test data with latest updates and modifications are the key elements determines the test project quality. Test data needs to be prepared when the test requirements are updated and available for the next phases of mobile testing.

Test Design

Test Scenarios: High level test scenarios need to be designed in line with test requirements and test use cases. 100% test coverage should be assured through the test scenarios. Test coverage tools like orthogonal array can be used to arrive at complete test coverage. Reference: Advanced Test Strategy

Test Cases: Each test scenario should result in one or many relevant test cases for the particular cycle. Test case should follow the standards as such any new test engineer should be in a position to read and understand the test steps.

Test Execution

Tests should be performed in respective test environment (mobile applications/emulators) to make sure that mobile applications are behaving as expected. If there is any discrepancy between actual results and expected results, new defects need to be raised to fix the issues.

Test Report

Execution report describes the test results and outcome of the testing cycles after the final phase of software test execution. Each test report has the respective authorities to sign off the software test reports as the report evidence the stability of the mobile applications to deploy in the real time environments.

Quality Parameters Tested

Correctness
E.g. Testing the mobile email messages to make sure that the correct email account is accessed.

Reliability
E.g. Automation Testing on Mobile Chat Application to check delays, inconvenience and difficulties in accessing the application functionalities

Usability
E.g. How user friendly the mobile is when the snapshot is captured and how easy it is to transfer to other devices in usable document format like JPG?

Maintainability
E.g. How easy to install multiple game applications and uninstall?

Reusability
E.g. How easy the reinstallation of the Operating System, features into mobile?

Testability
E.g. How practical it is in observing a reproducible functional flow of such application feature if they do exist in mobile?

Negative Impact if the Mobile Testing is Incomplete

If the mobile application quality is bad, it impacts the brand value of the organization and stakeholders. Negative product image is possible to spread in the product sales market. Brand loyalty will get decreased and it impacts the mobile purchase over the period. Brand equity improvement is impossible when there are production defects exist after mobile application testing

Types of Mobile Application Testing

Mobile Functionality Testing
Mobile Performance Testing
Mobile Usability Testing
Mobile Security Testing
Mobile Compatibility Testing
Mobile Interrupt Testing
Mobile Interoperability Testing
Mobile Localization Testing

Mobile Application Testing Tools in the Testing Industry

Experitest[4]

Widely used mobile application tool which facilitates downloadable plug & play set up and cross platform portability. Seven out of ten international banks are tested mobile applications using experitest products.

Perfecto Mobile[5]

Cloud-based solution for testing, measuring & monitoring mobile application quality using real devices and carriers

uTest[6]

uTest Tools facilitate excellent bug tracking system, test case results management, issues and event alerts as part of mobile tests

Neotys[7]

Widely used Mobile Performance Testing tool (NeoLoad) which allows to simulate mobile networks (3G, 3G+, H+, 4G LTE etc) from the Cloud or on-premise with bandwidth constraints, latency and packet loss simulation.

Soasta

Mobile Automation, Performance Testing and Real User Monitoring are some of the solutions from Soasta mPulse[8]

Robotium

Similar to Selenium, Robotium[9] used for Android OS based mobile testing supports Android features such as activities, toasts, menus and context menus.

Ranorex[10]

Object recognition method is recognized for mobile application testing based on Android, iOS and Windows 8 based mobile testing.

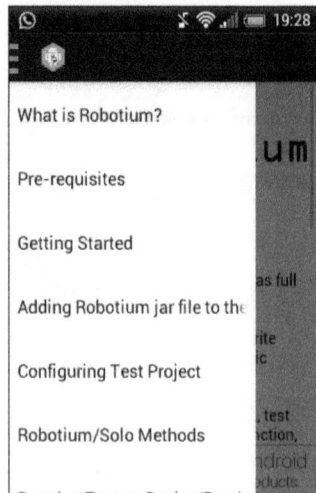

Figure 5 Robotium Screen

Eggplant

TestPlant 's eggOn[11] mobile testing tool allows mobile testing across any device and mobile OS without jail breaking.

Connection Modes for Mobile Testing

As part of test execution of the mobile app, mobile devices are connected with computers using USB or Wi-Fi, based on the mobile testable requirements. As part of negative testing.

Mobile Application Test Methods

Mobile Testing can be accomplished using Real Device or Emulators to test the mobile features.

Real Devices: USB or Wi-Fi is used to connect the real mobile devices in test execution.

Emulators: Software Imitates the functionality of real devices on computer which is different from real devices but the emulated behavior closely imitate the exact behavior of the real device and applications.

Real Device based Mobile Application Testing

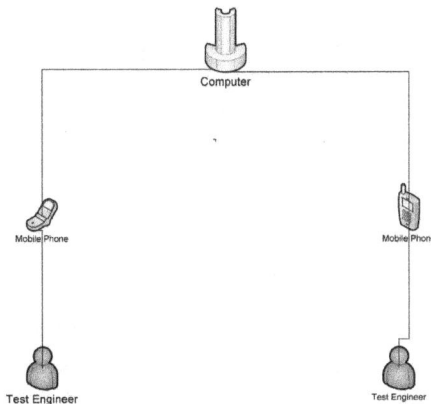

Figure 6 Mobile Device Setup

Configure the mobile set up using USB/Wi-Fi connection and install device drivers which are required for application access. After successful installation of device drivers, download the binary files which are required for application. Install the binary files in the mobile device using Application Manager or directly on the mobile device.

Open the installed application and perform smoke tests to prove that the application is ready for mobile testing.

Perform the test case steps and after successful test execution and defect retest, close the application and include the observations in 'End of Test Reports'.

Advantages of Mobile Application Tests

Reliability: Accurate Test Results.

Interoperability Testing: Real Mobile Device Testing is performed in live network to prove the interoperability of the devices.

Better User Experience: User friendliness of mobile device access is targeted in user experience testing.

Performance: Load and Stress based bottlenecks are identified and fixed during performance mobile tests which gives the better results over the mobile reliability.

Challenges in Mobile Application Tests

Expensive Logistics and Operations Management: On an average forty devices are tested in the mobile testing industry to prove the application readiness. One third of real mobile devices are replaced between the cycles due to updated versions of hardware and software. Procurement takes a major role in managing the logistics over mobile application tests.

Difficulty in design and development: Debugging process is very slow in initial stages of devices testing as real handsets are harder to connect to IDE than emulators.

Security Challenges: Mobile devices can be stolen very easily when it is connected in local network of workstations.

Emulators based Mobile Application Testing

Latest version of emulator can be downloaded and installed in the computer. Once installed, emulator needs to be connected to testing tool. After selection of Emulator type in the devices listed, it is flexible to operate mobile applications using computer keyboard and mouse hence emulator plays the key role of mobile devices in mobile testing.

Advantages of Emulator based Mobile Application Testing

Free of Cost: Most of the mobile device emulators are free for download and use. Also emulators are provided along with SDK as part of new Operating System releases.

Ease of use: Emulators' simplicity to access application by simply download, install and access is easy comparing to expensive real devices based testing.

Speedy test completion: When real devices are to be connected to local network and tested, emulators are simply to download, install and use mode as it is simple and speed in test completion.

Challenges of opting Emulators for Mobile Application Testing

Emulators doesn't provide the real mobile and real platform environment hence it won't create the reality of how the mobiles are used.

Interoperability Tests are not possible in emulators as users can't attend incoming calls, text messages in emulators like real devices. Also HSPDA,WCDMA,UMTS and LTE network technologies cannot be tested using emulators.

When emulators are used for testing, the memory usage and processing power won't be matching to the real mobile devices. So majority of the hardware and performance related issues are found only using real devices testing.

Emulators are connected to computers which has LAN based networks with corporate firewalls where as mobile devices are connected through radio interfaces; Moreover OS specific hardware and software requirements are not updated with Emulators where as when there is a new release in OS, mobile devices are replaced with latest versions for the better compatibility.

Mobile Application Testing Identification Techniques

Text: Identified by Optical character recognition (OCR) feature based on customized text identifying algorithms.

Native: Object identification is unique using attributes like "id", "name".

Image: Test objects are converted to images and matches with the run time images of GUI.

Web: Document Object Model (DOM) property is used. HTML DOM Nodes are widely used to test the documents, html elements and html attributes.

Prioritization between Manual and Automation Mobile Testing

When a new feature is introduced and new released, it is feasible to opt for manual testing where as regression tests and complex tests can be automated based on Return On Investment (ROI) analysis. If it is One Time Test (OTT), it is suggested to perform manual eye ball test as automation tests are time consuming and expensive for single test execution.

Importance of Mobile Application Testing

Quality validation and fault identification is done as part of Mobile Testing. The importance of doing mobile application testing is to get the defect free product as an end result. But the exhaustive testing is not possible at all the testing cycles.

Following are the suggested test cycles for mobile devices/emulators based testing:

Test Cycle	Mobile Devices		Emulators	
	Manual Testing	Automation Testing	Manual Testing	Automation Testing
Unit Test	Suggested			
System Test	Suggested			
System Integration Test				Suggested
Regression Test		Suggested		
Performance Test		Suggested		
User Acceptance Test	Suggested			
Security Testing		Suggested		
Compatibility Testing		Suggested		
GUI Testing		Suggested		
Synchronization Testing	Suggested			

Challenges in Mobile Testing Industry

QWERTY, Touch and Normal type input methods are the basic three types of mobile phones. Apart from these three, different screen sizes and software customized mobiles are available in the industry. Wide variety of mobile platforms/OS types are also a challenge in mobile testing. Mobile operators varies based on the region of customers and network infrastructure has its own limitations across the borders. In testing cycles, single scripting doesn't fulfill the requirements as keystrokes, input methods and menu structure are different from mobile to mobile! So structured Test Automation Frameworks and customized reusable scripts can solve the problems to the possible extent in testing cycles. Better option to opt when various devices involved in testing is emulators as it is cost effective. In network testing, concentrating only on LAN or Internet based tests by avoiding the lower layers can help in test execution. Also pay as you go options like opting the remote access to mobile applications can reduce the challenges to major extent for need based testing!

Mobile Performance Testing

Performance tests are generally executed to determine how a mobile or mobile applications performs in terms of responsiveness and stability under a particular work load. Connection speed and mobile accessibility are the key factors while testing mobiles as part of performance testing. Load testing is primarily concerned with testing that the mobile can continue to operate under a specific load, whether that be large quantities of data or a large number of users. This is generally understood as mobile scalability. Memory Usage, CPU Usage, Batter Consumption, Load on Servers under various conditions are monitored.

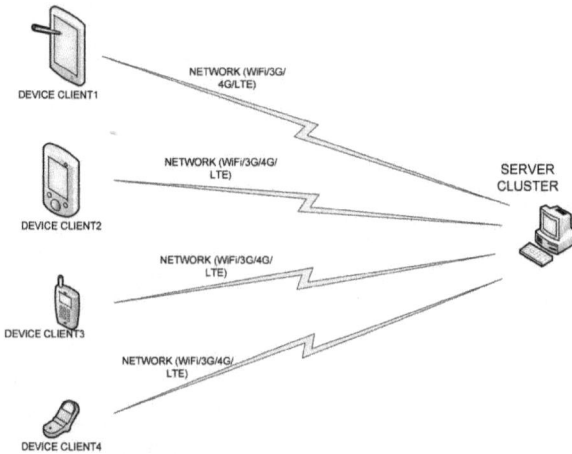

Figure 7 Mobile Performance Testing -Overview

Mobile Performance Test-Commonly Used Test Scenarios:
Low Battery Tests.
3G,4G,LTE Network Speed Tests.
No Network Available Tests.
Heavy Load Tests.
Battery Consumption Tests.
OS Based Performance Tests.
Response Time Tests.
Memory and Processor Utilization Tests.
Binary Size Native Application Tests.

Types of Performance Tests:

Client Side Mobile Performance Tests: Concentrating on the mobile as a client, mobile applications and Hardware usage.

Server Side Mobile Performance Tests: Concentrating on Application Server and Database Server usage.

Mobile Application Performance Testing Metrics:

Server-Monitoring Parameters: CPU Usage, Load, Bytes total, Process Time, User time, Packets sent/received.

Network-Monitoring Parameters: Packets and bytes sent/received, Average delay, Packet drops

Device-Monitoring Parameters: CPU and memory usage, Method level profiling, Web application component level performance, Response Time

Transaction: Response Time, Throughput

Cross Browser Testing

Web application testing in multiple browsers which checks application compatibility and correctness in different browsers. Client side and server side behavior of application is tested in mobile cross browser testing.

Open Source Tools used:Spoon Browser Sandbox, Browsershots, IE NetRenderer, Microsoft SuperPreview.

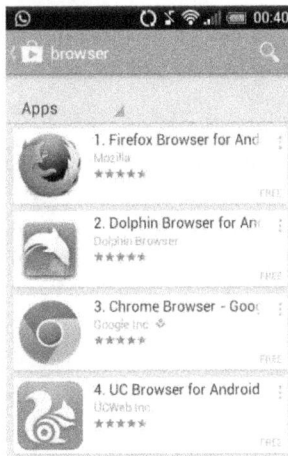

Figure 8 Browsers Search in Android Mobile

License Based Tools:Browsera, Adobe BrowserLab, BrowserCam.

Browsers used for Cross Browser Testing

Android OS-Nexus 7/4.2:Chrome Mobile 18,Dolphin Mobile 9.3,Firefox Mobile 19, Maxthon Mobile 4,Opera Mobile 12,Sleipnir Mobile 2.9.

Android OS-Galaxy S3/4.1: Android Browser 4.1,Chrome Mobile 27,Dolphin Mobile 9.4,Firefox Mobile 21, Maxthon Mobile 4, Opera Mobile 14, Sleipnir Mobile 2.10.

IOS-IPhone 5/6.0 Simulator: Mobile Safari 6.0 Browser

IOS-IPhone 4s/5.0 Simulator: Mobile Safari 5.1 Browser

BlackBerry OS-Blackberry Bold 9900: Blackberry Browser 9900.

Windows OS-Windows Phone 8: Internet Explorer 10 Mobile.

Mobile Testing Automation Frameworks

Test Automation Framework:

Structured usage of coding standards, test data handling and object repository management to have better return on investment (ROI) over test automation is achieved by using customized mobile test automation frameworks.

Return on Investment:

Following are the calculations used to understand the benefit of Test Automation:

Automation Testing Cost = Price Of Mobile Equipments + Price of Software Emulators(Testing Tools) + Development Cost + Maintenance Cost + Test Execution Cost.

Manual Testing Cost = Development Cost + Maintenance Cost + Test Execution Cost.

ROI = (Manual Testing Cost - Automation Testing Cost)/Automation Testing Cost.

Advantages

Increase Code usability: Reuse of code subset components across the test cycle is the benefit as the small change in the functionality needs a very minimal change in reusable components.

Higher Portability: Reusable components and test set up is highly portable for different project environment changes. It increases ROI when the test run in regular regression tests.

Reduced Script Maintenance Cost: During the long regression test execution plans, it is easy to make simple changes in reusable scripts which will get reflected across the modules it has been called part of test automation.

Programmable: When the test scripts are made into reusable elements for different functionalities of mobile app, it can be picked based on the needs in later part of test execution. Also the code is highly programmable for the customized requirements based on the usage of all types if-else, do-while formats.

Repeatable and Reusable Test Data: Once the Reusable Components are made to call in the main test script, test data also can be repeated and reusable according to the execution need. It reduces the test file size and increases the test execution speed.

Reliable and Comprehensive: Structured components and reusable parameters bring reliable frame of automation functional groups. Regular reusable format of script content help in bringing customized and exclusive test results for the testing needs.

Increased Productivity: Test resource utilization over the period of regular test execution will be best used when automation framework is implemented for emulator based mobile tests. Increased effort utilization is possible when regression tests are designed as per the predetermined automation plans.

Cost Reduction: Comparing to mobile equipment based testing, emulators based test execution is cost effective and tester friendly as well. It reduces the project costs and maintenance expenditures overall.

Superior Application Quality: Well managed object repository management and coding standards determine high level test quality standards and faulty test execution result is very less over repeated tests.

Reportable Results: Customized test results benefiting to tailor made requirements are the primary benefits over mobile test automation. Based on the business need, test reports and evidences can be customized and delivered on time.

Why Mobile Test Automation is Required?

Test Cases: When the test cases are large in numbers and need sever test iterations to run, it is good to automate and maintain test automation framework for better ROI.

API Level Test Execution: Application Programming Interfaces Level Test Design requires repeated test efforts in the same emulator level test validation. Once automated, can be reused across the test cycles without any changes to the code snippet unless the functionality change is initiated.

Verification of UI Control: User Interfaces are the main elements in Functionality Tests as it impacts the user behavior directly. It is important to validate the UI control and measurement over the long term when there are frequent software updates.

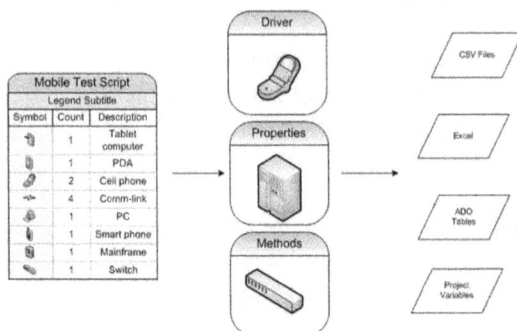

Figure 9 Mobile Test Script Overview

Handset specific features: Handheld device based testing requires handset based test codes to be executed based on each model's specification.

Automated Test Cases Structure:

Test Pack consists functional test cases refer to Drivers, Property and Methods which has the internal reference link to Excel, CSV,ADO table and Project Variables.

E.g.:

Test Case: Login to mobile application and navigate to help page

Design: Login, Navigation are split into different reusable components and used as a callable in main script to have better reusability of the test scripts. Login component refers to the CSV file which has the customer username and password encrypted.

Sample scripts from SeeTest

SeeTest generate the scripts in UFT (QTP), TestComplete, RFT, C#, JUnit, Python, Perl based scripting languages:

```
UFT (QTP) code snippet:
client.Launch "http://testwebsite.com/", true, true
Report
Java (JUnit 4) code snippet:
client.launch("http://testwebsite.com/", true, true);
Java (JUnit 3) code snippet:
client.launch("http://testwebsite.com/", true, true);
C# MSTest code snippet:
client.Launch("http://testwebsite.com/", true, true);
TestComplete code snippet:
client.Launch "http://testwebsite.com/", true, true
RFT code snippet:
client.launch("http://testwebsite.com/", true, true);
Python code snippet:
self.client.launch("http://testwebsite.com/", True, True)
Perl code snippet:
$client->launch("http://testwebsite.com/", 1, 1);
```

Data Driven Automation Mobile Framework

Test Script holds the logic of test case in this framework. Data is exclusively kept as excel sheets or different files apart from test script and used in different part of the scripts. External files like excel,csv,txt,odbc and ADO objects are used as data files in this framework. Test data is sourced from these external files and used as variable in the test script (variable: input values); Test library framework is widely used to efficiently use the less number of lines in the test scripts.

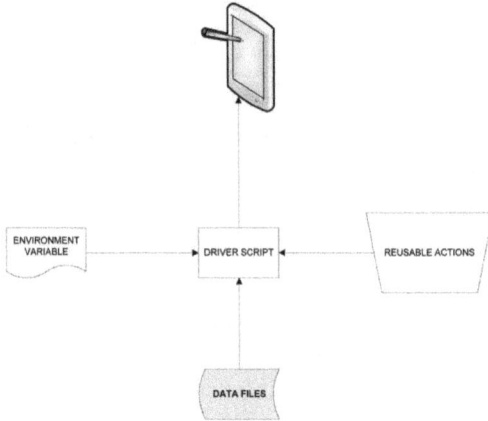

Figure 10 Data driven framework-an overview

Benefits

Test Script changes are minimal and test cases can be extended to test execution with multiple set of test data. High level of code reuse is possible and script maintenance is pretty much easier than other framework.

Challenges

Exceptional testing expertise and more amount of time required to design the data driven automation framework.

Keyword Driven Mobile Automation Framework

Data tables needs to be prepared and automation tools are required to call the data table and tests cases for test execution. Step by step approach is usually followed and emulators are used in most of the keyword driven frameworks to get faster test results. Keyword, Application Map and

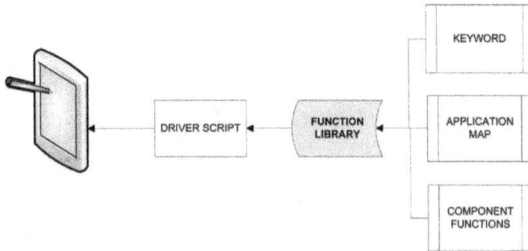

Figure 11 Keyword driven framework-an overview

Component Function are the three main features of the keyword driven mobile framework.

Keyword
Word acting as a key to the script. E.g.: St_Login.

Application Map
Object Repository are commonly known as application map which refers to the GUI components. E.g. Object Repository's Submit button object.

Component Function
Calling multiple variable in a function to achieve the particular process or functionality. Multiple components and multiple variable are used to create a better framework. E.g.: Login callable component used to login to application every time. Username and password are the variable used in this component.

Benefits
Keyword driven automation frameworks for mobile testing are highly reusable and the code has high opportunities to reuse since the tool is independent! Also the framework is independent of Application Under Test. Even script can be designed without AUT.

Challenges
Investment over the framework is quite high comparing to any other automation expenses. When the mobile application is big and complicated, this approach is followed in mobile testing projects. Test scripts are to be maintained with regular changes over a period of time. High automation skills set are required for this test approach.

Code Reusability

There is a benefit if we reuse the same set of developed test scripts across difference possible scripting languages using interfaces and abstracts. Coupling should be avoided to have better visibility of code reuse strategies.

Following are the programming languages used for mobile automation tools:

	Junit	C#	Perl	Python	Java Script	VB Script	VB.Net
ExperiTest-SeeTest	✓	✓	✓	✓	✓		
Perfecto Mobile						✓	
Robotium		✓					✓
Ranorex		✓					✓
Soasta					✓		
HP UFT						✓	

Responsiveness Testing of Mobile Applications

Responsive Web Design (RWD) is the method used to implement website with optimal viewing experiences. This technique reduces the extra amount of work in static website and it provides good compatibility over different device environments. So the website will be providing better response irrespective of device type. RWD testing is an emerging subsector in mobile application testing. Analyzing the response time and throughput variations across devices to access respective website is one of the primary test conducted in RWD tests. Aptus and RWD Bookmaklet are some of the best sourced tools used for RWD testing practice.

Aptus Studio

Based on Mac OS and used widely for responsiveness tests. It checks the adaptive behavior of website and custom screen size validations of applications. Load Agents are used in this tool to validate to device responses.

RWD Bookmarklet

iPad and iPhones are tested widely using this tool as the button which toggle on screen and keyboard are used to optimize the forms for iOS devices and test the requirements.

The Responsinator

Provides the mobile frame view of any website for difference devices like iPhone, iPad, Kindle, Tablet and other types of handheld devices. So the visibility of user experience is better comparing to any other tools used for frame based mobile website responsiveness testing.

ETVX Model

The Entry-Task-Validation-Exit model is used for mobile testing projects widely in the IT industry. If a particular activity failed in Mobile application, particular corrective activity is triggered as part of test execution. It can be used as part of development process. Each task can be classified into subtasks and further divided into subsets in ETVX Model.

Entry Criteria for Mobile Testing

Requirements should be delivered successfully to enter into the Testing Assignment. Stable mobile application should be ready for testing and test plan should have been reviewed and approved.

Task

Set of tasks to be performed as part of mobile testing like test scenario preparation, test case design, test data preparation, test case review, uploaded reviewed test pack to test execution tools as per the test execution activities.

Validation

Validation of the mobile products and application are taken care in this phase.
Validation: Are we testing the right product?'
Verification: Are we testing the product right?

Exit Criteria

Some of the major exit criteria are like relevant test cases should have been executed as part of test execution. Designed documents like test data, test results documents should have got approved. Test log and defect log are properly mapped to the test execution results. Review and sign off approvals of End of Test Report is one of the primary criteria to exit from the mobile testing projects

Constraints and Challenges over the Mobile Test Automation Projects

Mobile devices have been expanding across the platforms and browsers and rendering different offerings across the globe. Hence the device diversity is pretty much high comparing to any other product types, it requires an exclusive expertise over particular product type specialization. Expertise resource utilization over the latest technology and application is a major challenge in this automation projects. Multiple network operators and networks are used like GSM,GPRS,Wi-Fi, Wi-Max and connectivity speed vary across the region. So simulating the exact customer like scenario in the tests are bit challengeable in test automation. Because limitations over the processing speed and mobile memory size are tough to simulate in the emulators or the test environment.

Cloud Device based Mobile Testing:

In most of the big mobile testing projects, mobile emulators and applications are accessed through cloud devices by plugging the devices in mobile cradle point and connect it to host machine or cloud server. For this, cloud device tools should be installed on host location like 'Perfecto Mobile and SeeTest'. Also it is recommended to monitor the Application Server and Database Server memory usage as part of the mobile tests.

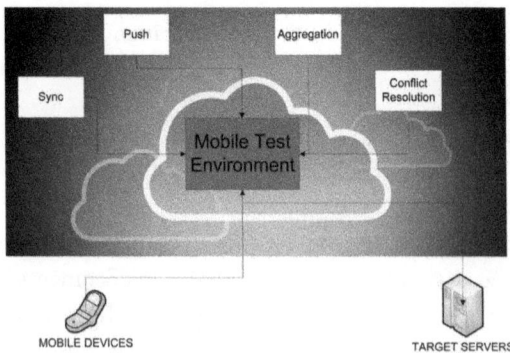

Figure 12 Cloud based mobile testing-an overview

Cloud based Mobile Testing-Test Procedure

As part of Mobile application testing, following are the common procedures used for testing each mobile device. As a first step cloud device tool needs to be installed successfully on host station. Once it is installed, cloud option should be selected. After cloud option selection, add the mobile device which needs to be tested. Add Host IP address and post to which mobile is getting connected for testing. Generate the test scripts to navigate to particular mobile application and test the required test items.

Benefits

High security is assured in Cloud based Mobile Testing as the application is handled from Cloud through Host Location. It is also easy to access the mobile application from cloud for testing purpose. No need of any proprietary hardware tools for the testing as the emulators are wide used for testing. Jail breaking and rooting are not need and resources can be utilized across the projects! Any mobile application can be run in local network for mobile testing cycles.

Challenges

It is also possible that any third party can access the files which are shared for testing. Cloud uptime can be one of the top project risk as the resources will be major impacted when environment is not available for test. When the mobile testing assignment is small, it will shoot up the storage and bandwidth cost of clouds used for the project.

Famous Mobile Application Automation Tools

Monkeyrunner: Common Script Based Tool
Robot Framework: White Box Testing Tool
Robotium: Blackbox Testing Tool

Famous Open Source Testing Tools for Mobile Testing

SIKULI: Used widely for automation testing of GUI using Image Recognition Technique for the test objects; The test script compare the actual snapshot against the expected snapshot image of the test item. Jython scripts

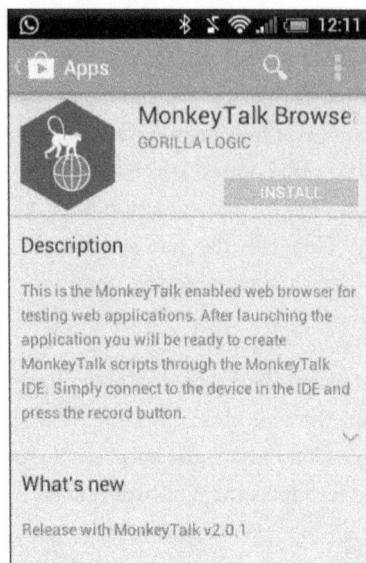

Figure 13 MonkeyTalk Browser Installation

are used in this tool and it runs in Windows, Linux and Mac platforms based mobile testing

Native Driver: API's UI will be implemented in Webdriver as part of Native Driver tests. User level interactions are same between the native and web based applications. Instrumentation applications are used in this tool.

MonkeyTalk: Used across the platforms as a functional testing tool. Some of the features are wide used in monkeytalk like script parameterization, Data driving, Expected result verification and screenshots. Javascript is supported in Monkeytalk

Appium: Used for Native and Mobile Applications testing across the platforms. Standard automation APIs are used in this tool and it supports Java, Objective C, Javascript, PHP, Python,Ruby,C#,Clojure or Perl with Selenium Web Driver for Mobile based application testing.

Frank: it is described as selenium for native iOS applications testing in mobiles. It uses cucumber and JSON to write structured tests. It includes Symbiote, which acts like an application inspector. When frank needs to be used with real devices, configuration needs to be modified and updated as per the mobile set up.

Robotium: Exclusive Test Cases are written using Robotium for Android Applications as part of black box and white box testing. It can also be used with JUnit to have the control over the complex tests.It gives the complete control over the Android SDK. Robotium install the test pack suit in an mobile device application in which test execution is performed.

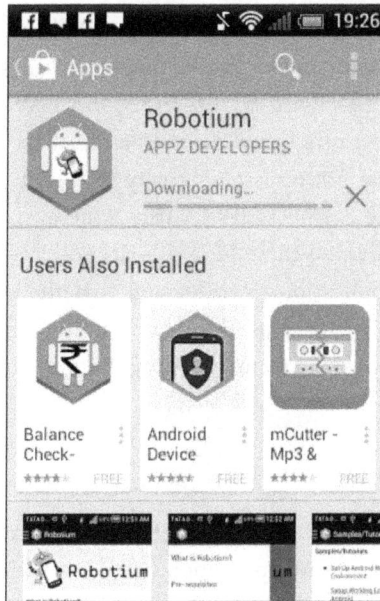

Figure 14 Robotium Installation

Calabash Android: Cucumber features can be executed in Android devices or the emulators assigned for test execution as part of mobile testing using Calabash Android. It creates cucumber skeleton in the present folder and run the tests. Also the cucumber features will have the .feature extension for the library files. The best part of the Calabash Android is Instrumentation (Test Server) as the installation of this tool covers the instrumentation in the application. The primary benefit is to use the latest version of the Calabash Android as it provides the predefined steps through step definitions folder.

Mobile Security Testing

Mobile devices and applications are the easy targets for the cyber attackers as the customer use from bank website to retail purchases are involving mobile usage. Especially certain automated script attacks are wide followed by attackers to various platforms of mobile devices including the protected mobile application packages.

Mobile Security Testing Basics:

The Open Web Application Security Project (OWASP) and SANS25 are some of the famous guidelines followed during security testing. Especially Top 10 Vulnerabilities listed by OWASP and CWE/SANS TOP 25 Most Dangerous Software Errors are the baselines for majority of the IT companies to set the priorities on security testing. E.g.: SQL Injection is one of the leading high priority security issue over the period of time in both the guidelines. So these high priority security issues needs to be tested in mobile application to make sure that the products are secure and protected from major vulnerabilities. 'Man in the Middle Attack' is a standard example proved in security testing that no proper solution found to overcome this particular attack![12]

Different Types of Mobile Security Issues

Malware Attacks: Google Play is widely used in many mobile formats. So users can download any application based on their need. It result in downloading malware software applications which prone to attack the mobile phones.

Case: 11000 Applications in Google's mobile marketplace have been affected and considered as malicious apps in 2011 and it has become 42000 apps in 2013.[13]

Recommendations: Anti Malware Apps will protect against mobile vulnerabilities.

Unauthorized Mobile Access: Personal contacts are stolen from mobile without users knowledge and used for spam and marketing for malicious attacks.

Recommendations: Users should be guided to avoid opening the strange emails with attachments or any links.

Encryption Issues: Clear text data transmission lead to data leaks and public key and private key encryption mechanisms are wide known for attackers. Automation scripts are used in malicious attacks to decrypt the private key transmissions.

Recommendations: Common encryption frameworks should be used with the standard guidelines to protect the mobile apps.

Data Leaks from Syncing: Reset links of webmail accounts like gmail or hotmail are prone for web attacks. When the data sync leads to data leak, it is wide used by attackers to capture the packet data transmission.

Recommendations: Users should be guided to use different password for each application and to change password frequently.

SQL Injection

Injection flaws, such as SQL script, OS, and LDAP injection occur when untrusted data is sent to an interpreter as part of a command or query (mostly).

E.g:
DB Injection-Mobile Test:
SELECT * FROM users WHERE id = '' OR '1'='1';
Instructions: When id is not null, the condition leads to '1'='1' which is true; So the complete list of data exported from database table 'user'.

Blind Injection-Mobile Browser Test:
http://www.testwebsite.com/test.php?id=sleep(30)
Instructions: If the sleep command is executed, the page will be loaded after 30 seconds. So any attacker can take advantage of sleep command

Error Message -Mobile Browser Test:
http://www.testwebsite.com/test.php?id='"
Instructions: If the error message is displayed for this URL, it prone to SQL injection attacks.

Tautology Injection-Mobile Browser Test:

http://www.testwebsite.com/test.php?username= ' or 1=1 /*&password=1

Instructions: If the username condition fails, condition leads to 1=1 which is true; so this condition enables the attacker login as administrator.

Broken Authentication and Session Management:

Incorrect implementations of authentication and session management leads to compromise of passwords, keys or session tokens, or to explore flaws to bring other user identities to attackers.

E.g:

Brute Force: Attackers try variety of passwords for the given username which is leaked through the mobile device; Automated malicious tests are conducted to simulate this normal brute force attacks.

Reverse Brute Force: Alternatively variety of user names are tried for the leaked password as a reverse brute force attack. This simulation can be automated by attackers to automation tools or addins.

Session Fixation Attack: Session of the victims can be transmitted to attackers using URL parameter or through the cookies in the mobile device. When the victim open the malicious website or the attacked url, session id is transmitted to attackers. When the cookie is stored in the browser, it can be transmitted to attacker through the specific malicious request. So the attacker uses the cookie or the session id in the request and access the respective web sources acting as a authorized victim.

Insufficient Session Expiration: When the user is just closed the website without logging out in the mobile, it can be accessed by the next user who access it. If the mobile memory is shared or the mobile is stolen, it can be possible to access the session ID to get the privilege of the victim.

Cross Site Scripting (XSS)

When application handles the untrusted data and send to web browser without validation leads to attackers to execute scripts in the victim's web browser which can hijack user sessions, deface websites or redirect to attack oriented websites.

Insecure Direct Object References

Direct object reference made by developer leads to exploring about internal directories, folder structures to attackers.

Security Misconfiguration

Incorrect Configuration over the application deployment, frameworks, application server, web server, database server and platform leads to this attack. Default configurations are often insecure and software updates needs to be installed time to time to avoid latest threats to the mobile devices.

Sensitive Data Exposure

Credit Card, Tax IDs and other authentic credentials are not protected in many websites which lead to data attack and hackers easily capture the sensitive information of whole stakeholders. Sensitive data needs critical protection through encryptions.

Missing Function Level Access Control

Access control checks are done usually at client level. But it has not been followed widely at server level. When the request is anonymously sent and accessed in server, it prone for server side attack due to missing access control.

Cross Site Request Forgery:

This attack forces the victim's browser to send http request and other authenticate information also shared along with the request to the respective application where attacker interested to receive information.

Using components with Known Vulnerabilities

When the fully liberated components are compromised, it is possible for the huge data loss through a subset of the code without victim's knowledge.

Unvalidated Redirects and Forwards

When untrusted data are used and untrusted webpages are accepted as default page, it leads to malicious redirects to malware websites.

Man In the Middle Attack[14]

A form of active eavesdropping in which the attacker decides to estabilish independent connections with the mobile and relays messages between the mobile user, making them believe that they are talking directly to each other over a private connection, when in fact the entire conversation is controlled by the attacker.

Social Engineering and Reverse Social Engineering

Social Engineering is a popular attack in present trend and there are no standard solutions to overcome from it. This section talks about different social engineering ways and steps to reverse engineering it to secure data.

Introduction- Social and Reverse Engineering-Present Approach:

Social engineering is the act of manipulating people into performing actions or divulging confidential information. Types: Phishing, IVR, Baiting, Pretexting, Diversion Theft and Quid pro quo. When the user is attacked by any of the Social engineering activities/processes, how to find precautions to resolve it? When attacks happen externally like network security hacks, it can be tracked by testing it externally. But when users are attacked using advanced techniques like 'Social Engineering', it should be tested internally. Applying Cognitive Psychology in Social Engineering can provide precautions to avoid hacks. The following are the types of social engineering, the normal flow(steps to hack) and Reverse Engineering Framework in which the social engineering attacks are tracked and reported.

Reverse Engineering Phishing mails:

Normal Flow:

a. Hacker sends Phishing URL in an email to user.

b. User opens the mail and click on the URL.

48

   c. The hidden script in the URL sends the username and password to hacker.

Reverse Engineering Framework:

   a. Hacker sends Phishing URL in an email to user.

   b. User opens the mail and identified that it could be a Phishing mail.

   c. User complaints to IC3 (Internet Crime Complaint Center) through Complaint Referral Form and contact NCIS (National Criminal Investigative Service)

   d. Police search the IP from which the mail is sent and find the ISP.

   e. Police call to ISP to provide the records of IP.(Since hacker changed his IP, the IP that the police have is not the current one so that is why they are asking hacker's ISP for records of the IP they have.)

   f. Police find collection of dynamic IPs and users.

   g. Investigate the latest IPs.

   h. Contact the Local Police to go home and investigate face to face.

   i. The police will take a quick look at hard drive using specialized tools and see if the person has ever had that IP that the police blacklisted. If it the IP is found, arrest the hacker.

Reverse Engineering of Baiting

Normal flow:

   a. In this attack, the attacker leaves a malware infected floppy disk, CD ROM, or USB flash drive in a location sure to be found in company premises.

   b. An unknowing employee might find it and subsequently insert the disk into a computer to satisfy their curiosity.

   c. The user would unknowingly install malware on it, likely giving an attacker unfettered access to the victim's PC and perhaps, the targeted company's internal computer network.

Reverse engineering framework:

   a. In this attack, the attacker leaves a malware infected floppy disk, CD ROM, or USB flash drive in a location sure to be found in company premises.

   b. An employee might find it and report to the security.

c. The organization can develop a policy and procedure to identify employees such as wearing a picture ID card. Visitors would be required to register and wear a temporary ID card. The employees should be encouraged to challenge anyone without a card.

Quid pro Quo

Normal Flow

a. The perpetrator will randomly call business numbers posing as technical support
b. Eventually they will stumble across someone who was waiting to be contacted by technical support
c. In the process of fixing an issue the perpetrator will have the victim execute commands that will give them access or control of the victim's system, install malware or have them divulge sensitive information (such as passwords) as part of a "quality survey"

Reverse engineering framework:

a. The perpetrator will randomly call business numbers posing as technical support
b. Eventually they will stumble across someone who was waiting to be contacted by technical support
c. Employee is aware of the company policy outside technical support and does not give control of the system/ any sensitive information to the perpetrator.

Phone Phishing:

Normal Flow:

a. The criminal either configures a war dialer to call phone numbers in a given region.
b. Typically, when the victim answers the call, an automated recording, often generated with a text to speech synthesizer, is played to alert the consumer that their credit card has had fraudulent activity or that their bank account has had unusual activity. The message instructs the consumer to call the following phone number immediately. The same phone number is often shown in the spoofed caller ID and given the same name as the financial company they are pretending to represent.

c. When the victim calls the number, it is answered by automated instructions to enter their credit card number or bank account number on the key pad.

d. Once the consumer enters their credit card number or bank account number, the visher (Phone Hacker) has the information necessary to make fraudulent use of the card or to access the account.

e. The call is often used to harvest additional details such as security PIN, expiration date, date of birth, etc.

Reverse engineering framework:

a. Don't give information to anybody unless you are certain you know who you're dealing with.

b. If you get a phone call about one of your accounts, hang up and call the institution.

c. Dial the number that appears on the back of your credit card or on your statements. Then, you know you're in the right place and they can take care of any issues on your account.

Pretexting:

Normal Flow:

a. A pretexter calls, claim he's from a survey firm, and ask you a few questions

b. When the pretexter has the information he wants, he uses it to call your financial institution.

c. He pretends to be you or someone with authorized access to your account. He might claim that he's forgotten his checkbook and needs information about his account.

d. In this way, the pretexter will be able to obtain personal information about you such as your SSN, bank and credit card account numbers, information in your credit report, and the existence and size of your savings and investment portfolios.

Reverse Engineering Pretexting:

a. Don't give out personal information like SSN, mother's maiden name, financial account numbers and other identifying information on the phone, through the mail or over the Internet unless you've initiated the contact or know who you're dealing with. And also legitimate organizations will have the information they need and will not ask you for it.

b. Be precautious. Ask your financial institutions for their policies about sharing your information and specifically about their policies to prevent pretexting.

c. Pay attention to your statement cycles. Follow up with your financial institutions if your statements don't arrive on time and review your statements carefully and promptly. Report any discrepancies to your institution immediately

d. Add passwords to your credit card, bank and phone accounts. Avoid using easily available information like your mother's maiden name, your birth date, the last four digits of your SSN or your phone number, or a series of consecutive numbers.

e. If you think that you are a victim report it to your financial institution immediately, Close accounts that have been tampered with and open new ones with new Personal Identification Numbers (PINs) and passwords.

f. Contact your local police as soon as possible, and ask to file a report. Even if the police can't catch the pretexter, having a police report can help you in clearing up your credit records later on.

Report the Social Engineering scams:

There are certain organizations are dedicated to investigate phishing as follows:

a. ConsumerFraudReporting.org.

b. Internet Crime Complaint Center (IC3).

c. Local FBI Office.

d. Contact your bank or Credit Card Company if you have given out credit card numbers, bank account information, etc.

For any specific scam/phishing, the respective authorities can be identified by 'ConsumerFraudReporting.org' website.

Phishing Report Analysis:

As per the 'Phishing Activity Trends Report-2010', 53% of computers are infected and 14% are related to Banking Trojan and Password. It explores that majority of phishing activities are related to banking and money transactions.

In October 2009, 26411 attacks are reported and it is slowly reduced in the remaining months. The average uptimes of phishing attacks has fallen steadily, and reached a notable low in 2009.

Affect of Phishing in Banking and Financial Sector:
Payment Services like Online Banking, Credit/ Debit Card Transactions are affected more comparing to other phishing attacks.

Use NLD forecasting technique

Combining more than two countermeasures may be a promising way to get more secure cryptosystem. The low sensitivity to the secret key and the potential possibility to mount attacks based on parameter estimation are regarded as the two greatest problems in almost all analog chaos based secure communication systems. Among all the known attacks, the ones based on and the NLD forecasting technique have received most attention, while little work has done on the low sensitivity to the secret key and the security problem of parameter estimation. It can figure out social engineering attacks and hacking possibility based on parameter estimations and security key analysis.

Use Best Protocols for transactions:
CAM can be used for mobile computers to inform their peers when their network address has changed; 802.1AE provides encapsulation and the cryptography framework for Ethernet protection.802.1af MAC Key Security supporting for key management, authentication and authorization along with 802.1AE. This is the best suitable to avoid Phishing in large transactions; For Wireless LANs, 802.11i is the best security protocol;

Use of best security Cryptography:
The best cryptography in the world will not protect a network that is otherwise vulnerable because of design faults or careless procedures that leave critical interfaces exposed to intruders. When selecting the cryptography or protocol, the following key points should be analyzed:

    a. Authentication and Non-repudiation Data Integrity

    b. Certification

    c. Data Protection

For all data deserving encryption, the following steps needs to be completed as per OWASP2010 guidelines: Considering the threats you plan to protect this data from (e.g., insider attack, external user), make sure you encrypt all such data at rest in a manner that defends against these threats.

    a. Ensure offsite backups are encrypted, but the keys are managed and backed up separately.

    b. Ensure appropriate strong standard algorithms and strong keys are used, and key management is in place.

    c. Ensure passwords are hashed with a strong standard algorithm and an appropriate salt is used.

    d. Ensure all keys and passwords are protected from unauthorized access

Conclusion:

Using secure protocols and systematic secured transactions over different layers of protocols can avoid Phishing activities. Social Engineering is a complex hacking method and securing data/transactions using reverse social engineering is essential for security and thus it is an emerging field which needs more research on it.

Categories in Mobile Security Testing:

Local files based tests, Browser settings tests, Application level tests, Network tests including https tests, Server based tests including auditing, authentication, authorization are some of the major tests performed on mobile devices as part of Mobile Security Testing Cycles.

Device Security Testing

Security measures are critically analyzed and tested for passwords, passphrases, unlock patterns, remote wipe tools and these measures are made as tests to prevent unauthorized parties from accessing the mobile devices.

Recommendations: Device security should be improved by using password or pin for each device as a mandatory criteria for access. Every period of inactivity, mobile device should be locked. Enabling the Wifi connectivity should be carefully monitored every time by the user. Application Installations should be reviewed and only trusted apps should be installed and used in mobile devices. Frequent releases of firm wares and software updates are very important for mobile devices to keep the mobile devices secure with the latest fixes against attacks.

Platform Security Tests for iOS based Mobiles

iOS have their own Application Interfaces and it has exclusive data protection features, iOS Keychain, firewalls and packet rules to protect the iOS platform. It's intrusion detection function provide the information about unauthorized access attempts over the mobile devices. As per the network usage plans, network security options can be chosen in iOS based mobiles. So the security level attacks are very less and the protection is high comparing to any other devices.

Network Level Mobile Security Testing

IP Spoofing: Any stranger sending messages to a host with an IP address (not the own IP address and usually it changes time to time) indicating that it has been sourced from a trusted host and gaining a specific unauthorized access to the host.

Routing (RIP) attacks: No built authentication given in RIP model. So any attacker can send the packet data just like the victim and automate in such a way that all the packets are sent from attacker's machine.

PING Flood (ICMP Flood): Attacker simply targets the victim's particular machine for particular target time and hit the requests like ICMP echo requests. So the system crash or slow down or the ping utilities cause the respective damages over the victim's system.

Teardrop Attack: Programs like Teardrop can be used to send IP fragments that can't be resembled properly by modifying the packet value and cause system reboot or hold the victim's respective set of systems.

Packet Sniffing: Any packet sniffer tool can be used by attacker to query the database or the hijack the account, password information from respective IP address.

Phishing: Sending lookalike of authentic emails and trying to provoke and get the information about the victim's confidential data through email. All personal information, financial information are gathered in this way and used for malicious attacks.

Recommendations to avoid mobile security attacks: Deploy firewalls for mobile devices and utilize IPsecVPN at the network layer;Mainly user authentication and data encryption needs to be used in data link layer; IPv6 with IPsec can be used for high level security for network layer.

Server Level Security Testing for Mobile Apps:

AAA Security: Authentication, Authorization, Accounting (AAA) are the focus as part of the server level security.

Suggested Protocols: As part of mobile security, RADIUS (RFC 2865) can be used to protect the apps. TACACS+ (developed by Cisco) also a best known to implement for mobile apps for better network security.

Widely Used Mobile Security Testing Tools:
HP Fortify,Whitehat Security, IBM Rational Appscan, Rapid7, Veracode, Kryptowire, Lint, Nessus

Security Tests-Requirements Framework
Confidential and secure information are hacked using automation tools like SQL inject me, Fiddler etc and most of the websites are hacked across the locations. Web security testing catches the path of hackers and tries to secure the applications in a smart way by vulnerability assessment. Requirements engineering is a first step which provides the clear picture on what to test.

Introduction- Requirements Engineering-Present Approach:

Most of the security testing requirements are written in generic business condition by Security Test Experts (STE). Signed off requirements are rarely revised and incorporated with updated security flaws. It leads to poor security testing on applications. Moreover 90% of security bugs are not linked to requirements due to less suitability. It clearly describes that 'Requirements' of Web Security Testing needs advanced 'Requirements Engineering' approach with respect to security flaws and risk analysis.

Why Requirements are important in Web Security Testing:

'What to test?' is important in functional testing requirements; 'How to test?' is very important with respect to Security Testing. The hackers ways needs to be tracked using 'Requirements'. Describe 'Requirements' is just a first step to test. But revise the requirements is a parallel approach which needs to be followed in every cycle of testing. Whenever new hacking techniques found/ major vulnerabilities tracked, it needs to be incorporated in upcoming cycles of 'Security Testing-Requirements'.

Security Test Life Cycle (SeTLC)
Prerequisites before SeTLC:

    a. Questionnaire Support Discovery
    b. Risk Profiling
    c. Attack Profiling

Test Life Cycle (SeTLC):

    a. Requirements Engineering
    b. Use cases generation/Scenarios
    c. Test Cases Design
    d. Execution
    e. Report Generation
    f. Defects (Defect Life Cycle)

Note: Requirements (with version number) should be linked to Use Cases/ Scenarios, Test Cases, Execution Cycle, Reports and Defect

Types of Defects in Security Test Life Cycle:

a. Suggestions
b. Feedbacks
c. Recommendations

Most of the accepted Suggestions, Feedbacks and Recommendations should be included as 'Requirements' in upcoming test cycles.

'Advanced Requirements Engineering'-Types of Requirements

a. OWASP Requirements
b. 'Inputs' Requirements
c. 'General Browser Settings' Requirements
d. Entry Point Test Requirements
e. Communication Point Test Requirements
f. Storage Point Test Requirements (Ex: Log forging Requirements)
g. Spoofing Requirement
h. Tampering Requirement
i. Repudiation Requirement
j. Information Disclosure Requirement
k. Denial of Service Requirement
l. Elevation of Privilege Requirement

OWASP Top10 vulnerabilities:

Security Test Requirements should be updated and incorporate with new updates of top 10 security flaws from OWASP releases. It makes 'Requirements phase' more updated and latest set of platform for security testing.

'Requirements' of OWASP (based on OWASP2007):

A1 – Cross Site Scripting (XSS) Requirement
A2 – Injection Flaws Requirement
A3 – Malicious File Execution Requirement
A4 – Insecure Direct Object Reference Requirement
A5 – Cross Site Request Forgery (CSRF) Requirement

A6 – Information Leakage and Improper Error Handling Requirement

A7 – Broken Authentication and Session Management Requirement

A8 – Insecure Cryptographic Storage Requirement

A9 – Insecure Communications Requirement

A10 – Failure to Restrict URL Access Requirement

'Inputs' as Security Test Requirements:

Inputs are the entry gates for hackers. Inputs are the major loophole for security flaws. Strong and smart validation/encryption of inputs leads to data safe and secure application. But, Security test should be done often on inputs based on latest flaws. So Defects of each cycle should be incorporated in Requirements of next cycles.

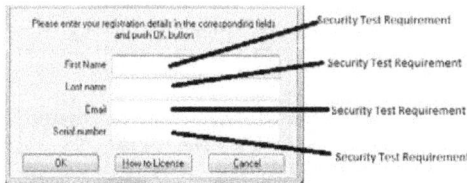

Figure 15 Security Test Requirements

Figure1: Consider each input as 'security test requirement' and validate it.

STE has to apply 'Parameterized Approach' to validate the inputs for requirements. Each input requirement has to update with types of valid inputs and types of invalid inputs.

'General Browser Settings' Requirements:

In general, web applications use java script for client side validation. It can be disabled and skipped easily by browser settings change. So strong validation needs to be implemented and it should not be overridden in any way. It should be added as a primary requirement and test with all the possibilities (use orthogonal array) of settings change.

'Entry Point Test' Requirements

Login page of web applications can be manipulated by Xmlhttprequest using JavaScript code. Similar to login page, the complete entry points of application need to be validated by adding security requirements.

'Communication Point Test'

Requirements:

HTTP Response Codes displayed whenever there is an error in displaying websites. These codes are well known and traceable by hackers. So each response code should be added as a security requirement and tested properly. As a best practice, same error message should be displayed for all the HTTP response codes.

'Storage Point Test' Requirements:

Back-end location points, log entries can be injected nicely without administrator's notice. In 'Log forging', it is simple to manipulate the log entry.

Actual log entry (when input given as 'abcd'):

Error: The user has entered a value that can't be passed as a number –abcd

Hacker's input:
\\n INFO: Admin has changed the password successfully

Now the log entry is manipulated as follows (manipulated log entry):

Error: The user has entered a value that can't be passed as a number –abcd
INFO: Admin has changed the password successfully.
Recommended action:
Requirements should be added with respect to log entries, log locations and log forging techniques.

Generic Security Test Life Cycle:

    a. Security Objectives and Requirements
    b. Security Design Guidelines
    c. Threat modeling
    d. Security architecture and design review
    e. Security code review
    f. Security testing
    g. Security deployment review

Requirements Engineering –Life Cycle:

    a. Questionnaire Support Discovery (QSD)
    b. Risk Profiling

    c. Attack Profiling

    d. Requirements

    e. Use cases/Scenarios

    f. Test Cases

    g. Execution

    h. Report Generation

    i. Defects (Recommendations, Suggestions and Feedbacks)

Requirements should be linked to use cases, test cases, execution cycle, report and defects. Recommendations, suggestions and feedbacks should be included as Questionnaire in 'QSD' of upcoming test cycles.

Tools based requirements:

Top most using tools of hackers needs to be listed based on our security testing and the corresponding requirements should be added in test cycles.

Fiddler Test Requirements:

Fiddler is a famous tool for HTTP validation in which complete HTTP requests and responses can be listed. So STE can perform following steps:

    a. Start Fiddler.

    b. Run the web application.

    c. Take the requests list from fiddler and write requirements.

'Latest Cracking Techniques' as Requirements:

In every test cycle, latest hacking techniques needs to be updated in test requirements and test cases. These updated hackings can be found in OWASP and other websites.

Engineering mechanism over the requirements:

Every requirement has to be maintained with version number and updated date. All requirements need to be revised by STE in consecutive test cycles. In line with requirements update, STE has to create and updated user stories for the respective requirements. It gives clear idea to create test cases combination.

Link Security Test Defects with Risk Profiling:

Every defect over the security test is called 'important asset' of test practice. Each defect needs to be analyzed and profile it along with risks. This risk profiles can be used across different projects to keep track on high risk factors and linked defects which are already reported as suggestions, recommendations and feedbacks.

Usage of attack profiles in the test cycles:

Comparing to risk profiles, attack profiles are unique and highly important for future security tests. Attack profiles can be linked to different 'Project Releases' done in specific attacks.

Update test wares in chain method: When an attack is found it should be updated and respective test wares should be generated parallel. So impact of Attacks/Vulnerability to Return On Invest (ROI) is manageable.

Attack and Recommendations:

When an attack is found and tested in the Application Under Test (AUT), respective 'Recommendations' are raised as new defects. These 'Recommendations' are the key solutions for the attacks hence all related recommendations should be linked to specific attacks in general. So when an attack is considered in specific test cycle, tester can refer related recommendations and it's fix status.

Required parameters for Requirements:

    a. Preferred number of attacks to test.

    b. Number of tabs to use for testing.

    c. Pause between adding items to sidebar (in ms).

The above parameters will be helpful when the security test needs to be done using Add-ins or open source tools.

SQL Injection Strings in Requirements

Recommended strings for SQL injection can be updated in test requirements. It can be taken by suggestions of previous test cycles. So 100% coverage on SQL injection strings is possible in all the test cases.

Validate 'Result Strings' and update the requirements

Hackers can easily track the database tables of the application using SQL injection. If the invalid input is inserted, application displays error message. For Example: select %u was reduced due to optimization. This message can be a hint to hackers and they can trace the backend tables using these

Conclusion

Requirements are the foundation for test cycles. Applying various engineering techniques over the requirements will give smart way to test security issues. Security testers are professional hackers hence security requirements are roadway to track all the vulnerabilities of AUT by experts. The list of requirement types provided in this paper can help STE to generate well established 'Requirements framework' for web security testing.

Defect Process Flow

IEEE Standard Classification for Software Anomalies (1044-1993) describes the details of defect management.

Defect Recognition

Recording the recognition:

When an anomaly is recognized in mobile application, supporting test input, script, screenshot and other data items are recorded to identify the anomaly

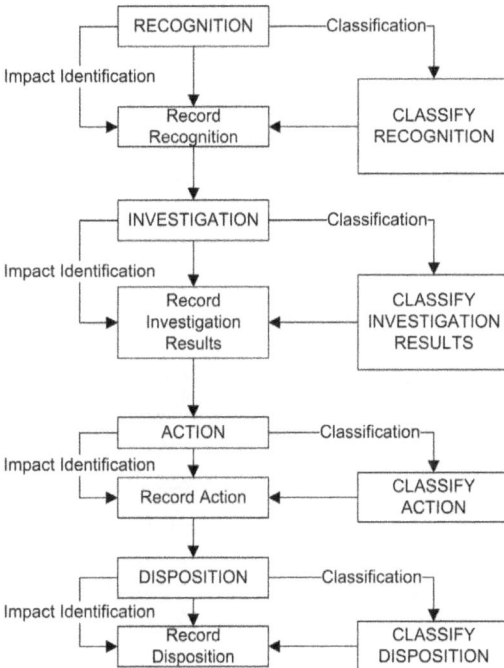

Figure 16 Defect Process Standards

and the environment in which it occurred like operating system, memory, and device type.

Classifying the recognition:

When the issue is found it is classified under analysis, review, audit, inspection, code/compile/assemble, testing, validation, support or walkthrough.

Identifying the impact:Defect owner (who raised the defect) describes about the impact of the issue in the mobile application functionalities like Urgent, High, Medium, Low or None. E.g. If the defect leads to mobile operating system crash or the mobile application crash it is classified as Urgent

Defect Investigation

Recording the investigation:

Supporting evidences are captured during investigation and new information are updated about the defect with reasons to it.

Acknowledgement	Verification
Date Received	Source of defect
Report number assigned	Data from recognition process
Investigator	
Name	
Code or functional area	
Email address	
Phone number	
Estimated start date of investigation	
Estimated complete date of investigation	
Actual start date of investigation	
Actual complete date of investigation	
Person hours	
Date receipt acknowledgement	
Documents used in investigation	
Name	
Revision	
Mobile Application Name	
Operating System	
Mobile Device	

Classifying the investigation:
During the investigation, defect is classified like Platform related, Hardware related with description. Also classification provided during recognition can be updated to the relevant type.

Identifying impact:
Previous impact description can be reviewed and updated for the respective mobile application based defects.

Action

Recording the action:
Defect fix details should be recorded properly for future reference as the same defect may occur in any of the next releases in which the below details can be referred.

Resolution Identification	Resolution Action	Corrective Action
Item to be fixed	Date resolution complete	Standards/Policies/ Procedures to be revised
Name	Mobile Device	Organization assigned corrective action follow-up
ID Number	Operating System	Person assigned corrective action follow-up
Revision	Version/Revision level	Correction action report number
Component within item	Organisation assigned to verify resolution	
Name	Person assigned resolution	
ID Number		
Revision		
Test Environment		
Mobile application name		
Test describing fix		
Planned date for action item's completion		
Person assigned action items		

Classifying the action:
Based on the fix category, action classification can be updated like software fix, test ware fix, operator training.

Identifying impact:
Based on project quality, reliability, risk, cost, schedule, mission safety and customer value, impact analysed on the defect fix as Urgent, High, Medium, Low or None.

Disposition

When the entire defect fix has been completed or identified to fix in long term, each defect can be disposed of by:

Recording the disposition:
Supporting evidences has to be collected while disposing the defect/anomaly as it can be used at the time of deployment and next releases.

Defect Disposition	Verification
Action implemented	Name
Date report closed	Date
Date document update complete	Version/Revision level
Customer notified	Method
Person sending notice	Test case
Date	
Type of notice	
Reference document number	

Classifying disposition:
When the defect is closed it can be classified like resolution implemented, not a problem, not in scope of project, outside vendor's problem, duplicate problem, deferred, merged with another problem or referred to another project (with reference)

Identifying impact:
Previously recorded impact category can be reviewed and updated while closing the defect as it is important to see the business and technical impact of the incident/defect while deploying to real time environment.

Defect Life Cycle

New

Test engineer finds the defect in mobile application and updates the defect portal (or tool) with defect description. This defect is yet to be assigned for analysis. It is important to state whether the defect is reproducible or intermittent.

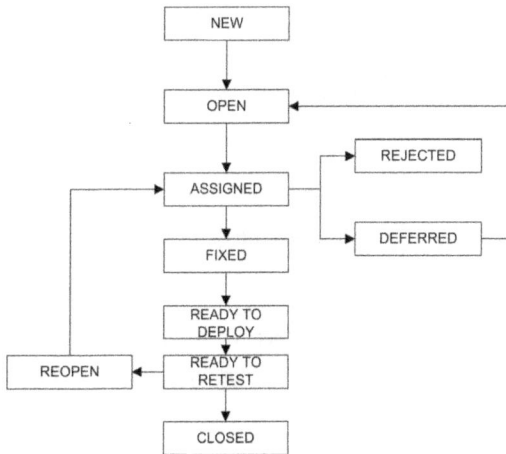

Figure 17 Defect Life Cycle

Reproducible defect: defect which can be found as per the particular set of steps when followed.

Intermittent defect: No proper steps or scenario found to simulate this defect again. But it occurs sporadically (known as sporadic behavior/sporadic defect).

Open

When the defect is raised with New status, It can be Opened and analyzed further to assign to respective developer.

Assigned

Defect assigned to respective developer who can fix the issue and it is developer's responsibility to update the defect status to fixed or deferred or rejected after fixing the defect.

Fixed

Once the defect is fixed and updated by developer, it can be changed to ready to deploy status

Ready to deploy:

When the build is ready for deployment in particular test environment or mobile devices, the status can be changed to ready to deploy.

Ready to test

When the fixed code is deployed and ready for testing team, it can be changed to the status of ready to retest.

Closed

Once the defect is retested after fix and found working as expected, status can be changed to 'closed' to dispose the defect.

Reopen

During the defect retest, if the defect found as not fixed or it leads to any other unexpected deviation or the incident, the defect can be reopened and assigned back to developer for further investigation and defect re-fix.

Deferred

When the defect is found not urgent or due to business reasons, needs to be fixed in further releases, it can be deferred and fixed in the later releases. So when the defect is updated as deferred it has to be updated with the comments on which version it is going to be fixed.

Rejected

If the defect is irrelevant or the duplicated defect due to earlier defects raised or any other valid reasons, it can be rejected and updated with valid comments.

Sample Defect Life Cycle of a defect

Stage1

Date Raised	: 20/05/2017
Priority	: High
Impact Severity	: Major

Reproducible	: Yes
Detection Phase	: System Test
Raised By	: Test Engineer1
Assigned to	: Test Lead
Status	**: New**
Target Cycle	: Cycle1
Impact	: 10000 CAD
Risk Value	: 2000000 CAD
Escalate	: No

Description

Subject	: Defect in hellowapp-login page
Steps to reproduce	:

1. Switch on the mobile device;
2. Launch the hellowapp mobile application;
3. Click on login button;
4. Enter valid user name and password;
5. Click on login button

Expected result	: User should login to hellowapp
Actual result	: Error Occurred message displayed
Attachments	: Error.jpg
Linked Entities	: Hellowapp Login Test Case1
Status Update	: NA

Stage2

Date Raised	: 20/05/2017
Priority	: High
Impact Severity	: Major
Reproducible	: Yes
Detection Phase	: System Test
Raised By	: Test Engineer1
Assigned to	: Development Manager1
Status	**: Open**
Target Cycle	: Cycle1
Impact	: 10000 CAD
Risk Value	: 2000000 CAD
Escalate	: No

Description

Subject	: Defect in hellowapp-login page
Steps to reproduce	:

    1. Switch on the mobile device;

    2. Launch the hellowapp mobile application;

    3. Click on login button;

    4. Enter valid user name and password;

    5. Click on login button

Expected result	: User should login to hellowapp
Actual result	: Error Occurred message displayed
Attachments	: Error.jpg
Linked Entities	: Hellowapp Login Test Case1
Status Update	:

'Assigned to development manager for verification on the code with the help of development team'-Test Lead

Stage3

Date Raised	: 20/05/2017
Priority	: High
Impact Severity	: Major
Reproducible	: Yes
Detection Phase	: System Test
Raised By	: Test Engineer1
Assigned to	: Development Engineer1
Status	**: Assigned**
Target Cycle	: Cycle1
Impact	: 10000 CAD
Risk Value	: 2000000 CAD
Escalate	: No

Description

Subject	: Defect in hellowapp-login page
Steps to reproduce	:

    1. Switch on the mobile device;

    2. Launch the hellowapp mobile application;

    3. Click on login button;

4. Enter valid user name and password;

5. Click on login button

Expected result	: User should login to helloapp
Actual result	: Error Occurred message displayed
Attachments	: Error.jpg
Linked Entities	: Helloapp Login Test Case1
Status Update	:

'Assigned to development manager for verification on the code with the help of development team'-Test Lead.

'Assigned to development engineer to analyze this issue in code'-Development Manager.

Stage 4

Date Raised	: 20/05/2017
Priority	: High
Impact Severity	: Major
Reproducible	: Yes
Detection Phase	: System Test
Raised By	: Test Engineer1
Assigned to	: Development Manager1
Status	**: Fixed**
Target Cycle	: Cycle1
Impact	: 10000 CAD
Risk Value	: 2000000 CAD
Escalate	: No

Description

Subject	: Defect in helloapp-login page
Steps to reproduce	:

1. Switch on the mobile device;

2. Launch the helloapp mobile application;

3. Click on login button;

4. Enter valid user name and password;

5. Click on login button

Expected result	: User should login to helloapp
Actual result	: Error Occurred message displayed

71

Attachments   : Error.jpg
Linked Entities   : Hellowapp Login Test Case1
Status Update   :
'Assigned to development manager for verification on the code with the help of development team'-Test Lead.
'Assigned to development engineer to analyze this issue in code'-Development Manager.
'Fixed in the build 3.0.9.0'-Development Engineer.

Stage 5

Date Raised   : 20/05/2017
Priority   : High
Impact Severity   : Major
Reproducible   : Yes
Detection Phase   : System Test
Raised By   : Test Engineer1
Assigned to   : Environment Manager1
Status   : **Ready to Deploy**
Target Cycle   : Cycle1
Impact   : 10000 CAD
Risk Value   : 2000000 CAD
Escalate   : No
Description
Subject   : Defect in hellowapp-login page
Steps to reproduce   :
                     1. Switch on the mobile device;
                     2. Launch the hellowapp mobile application;
                     3. Click on login button;
                     4. Enter valid user name and password;
                     5. Click on login button
Expected result   : User should login to hellowapp
Actual result   : Error Occurred message displayed
Attachments   : Error.jpg
Linked Entities   : Hellowapp Login Test Case1
Status Update   :

'Assigned to development manager for verification on the code with the help of development team'-Test Lead.

'Assigned to development engineer to analyze this issue in code'-Development Manager.

'Fixed in the build 3.0.9.0'-Development Engineer.

'Assigned to environment team for deployment of latest fix release 3.0.9'-Development Manager.

Stage 6

Date Raised	: 20/05/2017
Priority	: High
Impact Severity	: Major
Reproducible	: Yes
Detection Phase	: System Test
Raised By	: Test Engineer1
Assigned to	: Test Lead1
Status	**: Ready to Retest**
Target Cycle	: Cycle1
Impact	: 10000 CAD
Risk Value	: 2000000 CAD
Escalate	: No

Description

Subject   : Defect in hellowapp-login page

Steps to reproduce   :

1. Switch on the mobile device;
2. Launch the hellowapp mobile application;
3. Click on login button;
4. Enter valid user name and password;
5. Click on login button

Expected result	: User should login to hellowapp
Actual result	: Error Occurred message displayed
Attachments	: Error.jpg
Linked Entities	: Hellowapp Login Test Case1
Status Update	:

73

'Assigned to development manager for verification on the code with the help of development team'-Test Lead.

'Assigned to development engineer to analyze this issue in code'-Development Manager.

'Fixed in the build 3.0.9.0'-Development Engineer.

'Assigned to environment team for deployment of latest fix release 3.0.9'-Development Manager.

'build is deployed in latest test environment of mobile device 1b2 and emulator 34'-Environment Manager.

Stage7

Date Raised	: 20/05/2017
Priority	: High
Impact Severity	: Major
Reproducible	: Yes
Detection Phase	: System Test
Raised By	: Test Engineer1
Assigned to	: Test Engineer1
Status	**: Ready to Retest**
Target Cycle	: Cycle1
Impact	: 10000 CAD
Risk Value	: 2000000 CAD
Escalate	: No
Description	
Subject	: Defect in hellowapp-login page
Steps to reproduce	:

1. Switch on the mobile device;
2. Launch the hellowapp mobile application;
3. Click on login button;
4. Enter valid user name and password;
5. Click on login button

Expected result	: User should login to hellowapp
Actual result	: Error Occurred message displayed
Attachments	: Error.jpg
Linked Entities	: Hellowapp Login Test Case1

Status Update   :
'Assigned to development manager for verification on the code with the help of development team'-Test Lead.
'Assigned to development engineer to analyze this issue in code'-Development Manager.
'Fixed in the build 3.0.9.0'-Development Engineer.
'Assigned to environment team for deployment of latest fix release 3.0.9'-Development Manager.
'build is deployed in latest test environment of mobile device 1b2 and emulator 34'-Environment Manager.
'Assigned to test engineer for defect retest'-Test Lead.

Stage8

Date Raised   : 20/05/2017
Priority   : High
Impact Severity   : Major
Reproducible   : Yes
Detection Phase   : System Test
Raised By   : Test Engineer1
Assigned to   : Test Engineer1
Status   **: Closed**
Target Cycle   : Cycle1
Impact   : 10000 CAD
Risk Value   : 2000000 CAD
Escalate   : No
Description
Subject   : Defect in hellowapp-login page
Steps to reproduce   :
                       1. Switch on the mobile device;
                       2. Launch the hellowapp mobile application;
                       3. Click on login button;
                       4. Enter valid user name and password;
                       5. Click on login button
Expected result   : User should login to hellowapp
Actual result   : Error Occurred message display

Attachments   : Error.jpg, Test evidence.zip

Linked Entities   : Hellowapp Login Test Case1

Status Update   :

'Assigned to development manager for verification on the code with the help of development team'-Test Lead.

'Assigned to development engineer to analyze this issue in code'-Development Manager.

'Fixed in the build 3.0.9.0'-Development Engineer.

'Assigned to environment team for deployment of latest fix release 3.0.9'-Development Manager.

'build is deployed in latest test environment of mobile device 1b2 and emulator 34'-Environment Manager.

'Assigned to test engineer for defect retest'-Test Lead.

'This defect has been retested and found working as expected in the test environment device 1b32. Attached are the test evidences'-Test Engineer.

Sample Master Test Plan
Test Plan Identifier MAT-MTP.6.27

References
Requirement Specification Document v1.0.8.9

Introduction
This is the Master Test Plan for the Mobile Platform Message Service Project-Release 27. This plan will address only the items and elements related to message service project as part of release 27. The primary focus of this plan is to ensure that the new message services application provides the same level of information and detail as the current system while allowing for improvements and increases in data acquisition and level of details available.

The project will have five levels of testing, Unit, System, Integration, Performance and Acceptance. The details for each level are addressed in the approach section and will be further defined in the level specific plans

The estimated timeline for this project is very aggressive (Eight (8) months),as such, any delays in the development process or in the installation and verification of the third party software could have significant effects on the test plan. The acceptance testing is expected to take twenty days from the application delivery from system test and is to be done in parallel with the current application process.

Test Items
The following is a list, by version and release, of the items to be tested:
- a. Hello App, version 4.1
- b. HwsU Mobile App for Android, version 1.094
- c. Custom Mobile Configuration Package, version 1.0
- d. Phone Directory Updater-Automatic, version 9.13

Software Risk Issues
There are several parts of the project that are not within the control of this project but have direct impact on the process and must be checked as well.
- a. Local vendor supplied portable package for mobile networks. This package will be providing all the reformatting support to the mobile device versions.

b. Backup and Recovery of the packet transmission files-must be carefully checked

c. The ability to restart the mobile applications in the middle of the process is a critical factor to application reliability. This is an important factor in the cases of file sharing between the mobile devices.

Features To Be Tested

a. Applications of IOS-IPhone 4s/5.0 : Mobile Safari 5.1 Browser

b. Applications of IOS-IPhone 5/6.0 : Mobile Safari 6.0 Browser

Features Not To Be Tested

a. Network security and Service Layer Access

b. Hardware Assembly

Approach (Strategy)

Soasta is suggested for the Emulator based testing across the cycles followed by Eggplant tool for custom applications.

Eggplant required special training to test resources which will be delivered by mandatory training programs as per the plan.

ITIL standards should be followed for the environment and support related issues and high priority issues needs to be handled on time.

Cloud computing is the primary approach to share the emulators across the testing teams and users for acceptance testing.

Subset of regression test pack needs to be executed for every code drop and fix for the high priority defects.

Item Pass/Fail Criteria

Tests should be passed and completed only when test evidences are provided and reviewed by business.

Additional tests needs to performed when the tests are failed and Ready to Retest in order to pass the tests.

Subset of Regression Test Pack needs to be executed as and when the batch release for set of defect fixes on Failed Tests. So the Pass Criteria in this case is to prove that the entire module is working as expected after the defect fixes.

Suspension Criteria and Resumption Requirements

If the number of major defects reaches ten in particular test cycle, the test execution has to be stopped until the next build release as to follow on testing has no value and it makes no sense to continue the test.

Test Deliverable

Master Test Plan
Test Scenarios High Level
Test Cases
Test Data Sheet
Tools and their inputs
Emulators
Mobile Test Framework
Error Logs and Execution Logs
Problem Reports and Corrective Actions

Remaining Test Tasks

Task	Assigned To	Status
Create Test Data Sheet	Test Engineer	
Create Mobile Test Framework	Automation Tester	
Verify emulator prototype	Test Lead	
Verify Cloud Computing Environment	Environment Manager	
Verify the Mobile Test Resource Plan	Test Manager	

Environmental Needs

The following elements are required to support the overall test effort at all levels within the reassigned mobile project:

a. Access to both development and production based emulators and mobile applications
b. An exclusive test environment for performance testing as it should impact the regular test executions of other test cycles

Staffing and Training Needs

Preferred to have exclusive test resource for performance testing.

All testers need to be trained in Eggplant as it is essential to undergo custom based application training before test execution

Responsibilities
Test Execution Signoff: Test Manager
Test Evidences Signoff: Business Analyst
End of Test Report Sign off: Project Manager

Schedule
Test Cycle based schedule is updated in QC for Test Execution. Overall mobile testing assignment is schedule for nine months.

Planning Risks and Contingencies
    a. Limited Test Resources for Performance Testing
    b. Unavailability of Resources during Regional Holidays

Approvals
Project Sponsor:Michael Andrew
Development Management: Chidambraiah
Project Manager: Anurag Bhatacharya
Environment Manager: Deshmukh Pande

Sample Defects

#	CR/ Bug	Bug/Enhancement identified	Steps to replicate	Identified by	Date
1	Bug	User Name Visibility	While login in, user name is visible partially in the screen. Take a look at the screenshot please	Test Engineer1	\<Date\>
2	Bug	Incorrect email id leads to 'Error Occurred'	Login> Click on Forgot Password? Enter 'email@gmail. com' and submit it>It displayed Error Occurred	Test Engineer2	\<Date\>

Defect1&2 Test Evidence:

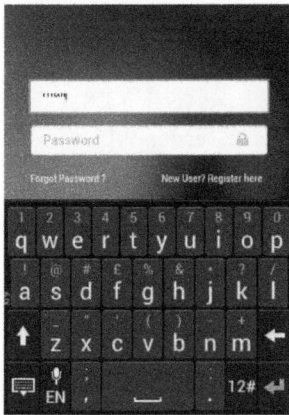

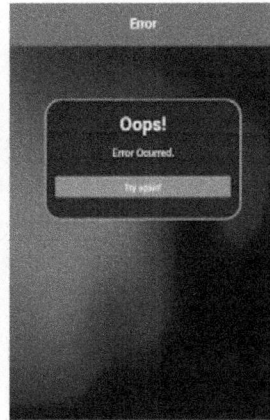

Sample Test Summary Report

Test Summary Report Identifier MAT-EOTR.6.27

Summary

This is the Test Summary Report for the Mobile Platform Message Service Project-Release 27. This report will address only the items and elements related to message service project as part of release 27. The primary focus of this report is to deliver the test results to ensure that the new message services application provides the same level of information and detail as the current system while allowing for improvements and increases in data acquisition and level of details available.

The report have five levels of testing results, Unit, System, Integration, Performance and Acceptance test results.

Test Items

The following is a list, by version and release, of the items which are tested:

    a. Hello App, version 4.1

    b. HwsU Mobile App for Android, version 1.094

    c. Custom Mobile Configuration Package, version 1.0

    d. Phone Directory Updater-Automatic, version 9.13

Environment

Test Environment version 2.0 is used for all cycles of this project.

References

MAT-MTP.6.27

Variances

Test Schedule is delayed by 2 days in Unit tests due to the Change Request 1.0.2.2

Comprehensive Assessment

Test Packs have been revised based on Risk based test approach to meet the deadlines and covered 100% functionalities and objectives as mentioned in Test Plan

Identification of Intermittent issues:

Defect ID 0034 and 0058 are raised based on Intermittent issues and 'Deferred' to analyze for future releases as both the defects are not reproducible and minor in nature.

Summary of Results

Test Cycle	Total Test Cases	Executed Test Cases	Passed	Failed	Defects	Test Completion Percentage
Unit Test	49	49	49	0	34	100%
System Test	38	38	38	0	12	100%
System Integration Test	15	15	15	0	4	100%
Regression Test	60	60	60	0	3	100%
Performance Test	15	15	15	0	1	100%
User Acceptance Test	39	39	39	0	2	100%

Evaluation

All major defects are fixed, retested, closed by respective test engineers and reviewed by business for the current release.

Limitations

Defect ID 0089 is deferred for implementations in the next release of HwsU Mobile App 1.095. So Search functionality in HwsU Mobile App will not be fixed in the present release and it has been agreed by business.

Summary of Activities

Test Schedule delayed in Unit Tests due to change requests and test pack has been revised in regression tests due to time delay of testing activities during system integration tests.

Approvals

Project Sponsor:Michael Andrew
Development Management: Chidambraiah
Project Manager: Anurag Bhatacharya
Environment Manager: Deshmukh Pande

Example Scripts for Mobile Test Automation using UFT (formerly HP QTP):

UFT (Unified Functional Testing) is widely used along with ALM (Application Lifecycle Management) for the testing projects in leading companies. Following are some of the common used functions for mobile automation scripting:

Wide Used Application Functions:

Application.Button.Click: Clicks on a button in an mobile application by calling the button control through Application Interface (API).

Application.Button.Info: Property parameter of a button can be used to specify the control property to retrieve value into defined variable.

Application.Text.Info: Property parameter of an information can be used to specify the control property to retrieve value into defined variable.

Application.Text.Click: Clicks on a control in an mobile application by calling the control through Application Interface (API)

Application.Image.Click: Clicks on an image in mobile application by calling the image control API.

Application.Image.Info: Retrieves the specified image control property in mobile application

Sample UFT Script:

DeviceAnywhere is one of the leading automation tool used for mobile automation along with UFT. Following are the sample scripts to understand the mobile automation. DeviceAnywhere API functions needs to be installed to practice the below scripts:

```
HWSUDeviceApi("HWSUDeviceObj1").InitiateObject 5582
HWSUDeviceApi("HWSUDeviceObj1").WaitForDevice 300
HWSUDeviceApi("HWSUDeviceObj1").SendKeys "[Touch(233,343)
[Home]","Beta"
HWSUDeviceApi("HWSUDeviceObj1").SetTextUseForeground True
HWSUDeviceApi("HWSUDeviceObj1").SetTextRGB 234,453,342
HWSUDeviceApi("HWSUDeviceObj1").SetTextThreshold 12
HWSUDeviceApi("HWSUDeviceObj1").WaitForTextPhone,31
HWSUDeviceApi("HWSUDeviceObj1").UnlockDevice
```

Instructions: 'InitiateObject 5582' initiates the test object in the test. '
WaitForDevice 300' waits for the device to be readily available for a specified
amount of time in seconds.

Finally 'Wait for Text Phone,31' Waits for for specified text to appear on the
device screen.

Index

Book References

[1]Diego Torres Milano, "Android Application Testing",PACKT Publishing

[2]Hung Q.Nguyen,Bob Johnson, Michael Hackett, "Testing Applications on the Web:Test Planning for Mobile and Internet-Based Systems,Second Edition",Wiley Publications

[3]Laurence T.Yang,Agustinus Borgy Waluyo Jianhua Ma,Ling Tan, Bala Srinivasan, "Mobile Intelligence", Wiley Publications

[4]Steven Splaine, "Testing Web Security", Wiley Publications

[5] Greg Fournier, "Essential Software Testing- A Use Case Approach"

Web References

1. www.scqaa-oc.com/wp-content/uploads/2013/01/Mobile-Trends.ppt
2. http://www.emarketer.com/Article/Smartphone-Users-Worldwide-Will-Total-175-Billion-2014/1010536
3. Kolawa, Adam; Huizinga, Dorota (2007). Automated Defect Prevention: Best Practices in Software Management. Wiley-IEEE Computer Society Press. p. 426. ISBN 0-470-04212-5.
4. http://experitest.com/
5. http://www.perfectomobile.com/articles/why-perfecto-mobile-0
6. https://www.utest.com/mobile-app-testing
7. http://www.neotys.com/introduction/mobile-load-testing.html
8. http://www.soasta.com/products/mpulse/
9. http://en.wikipedia.org/wiki/Robotium
10. http://www.ranorex.com/mobile-automation-testing.html
11. http://www.testplant.com/eggplant/testing-tools/
12. Blog: http://security.stackexchange.com/questions/6437/how-would-one-fully-protect-himself-against-man-in-the-middle-attacks
13. http://www.computerworld.in/news/malware-infected-android-apps-spike-in-the-google-play-store
14. http://en.wikipedia.org/wiki/Man-in-the-middle_attack

About the Author

Narayanan Palani, a leading digital technology specialist with considerable years of experience in software testing including research and development. Proven track record of success with seven international research papers, thirteen international certifications from globally acclaimed institutions such as the British Computer Society, PRINCE2, AXELOS, Hewlett-Packard, IBM and Wipro. Authored a book (Advanced Test Strategy published by Penguin Partridge Publishers) on digital technology which stands as one of the best reviewed books across the globe.

in uk.linkedin.com/in/narayananpalani/

www.ingramcontent.com/pod-product-compliance
Lightning Source LLC
Chambersburg PA
CBHW032013190326
41520CB00007B/461